ASIAPAC COMIC SERIES

WATER MARVIN

Marsh at the Foot of Liangshan

In Six Volumes:

Emergence of the Demons
Marsh at the Foot of Liangshan
The Tiger Slayer
The Siege of Zhu Family Village
Confronting the Royal Court
The Grand Assembly

Written by Shi Naian
Illustrated by Teo Seng Hock
Translated by Wu Jingyu

ASIAPAC • SINGAPORE

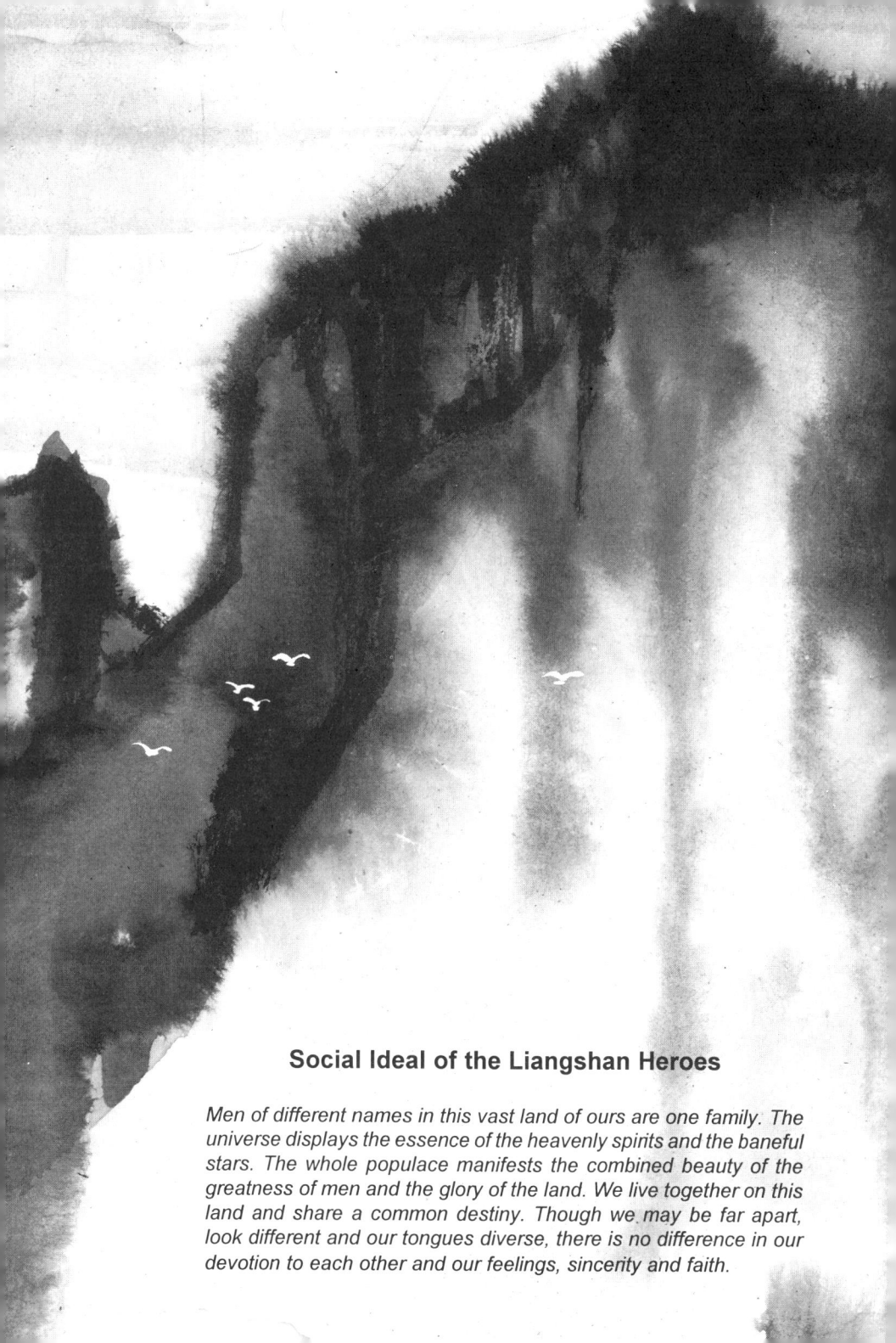

Social Ideal of the Liangshan Heroes

Men of different names in this vast land of ours are one family. The universe displays the essence of the heavenly spirits and the baneful stars. The whole populace manifests the combined beauty of the greatness of men and the glory of the land. We live together on this land and share a common destiny. Though we may be far apart, look different and our tongues diverse, there is no difference in our devotion to each other and our feelings, sincerity and faith.

Publisher
ASIAPAC BOOKS PTE LTD
996 Bendemeer Road #06-08/09
Kallang Basin Industrial Estate
Singapore 339944
Tel: (65) 392 8455
Fax: (65) 392 6455
Email apacbks@singnet.com.sg

Visit us at our Internet home page
www.asiapacbooks.com

First published August 1998

Cover illustration by Teo Seng Hock
Cover design by TAP DESIGN PTE LTD
Body text in 8/9 pt Helvetica
Printed in Malaysia by
Caxton Printing Sdn Bhd

Publisher's Note

WATER MARGIN is one of the four best-known Chinese literary classics, which also include *Journey to the West, Strange Tales of Liaozhai*, and *Dream of Red Mansions.*

Set in the Song Dynasty, the novel relates how 108 men and women gathered on Liangshan Mountain, which is today's Shandong Province in China. The heroes became leaders of an outlaw army of thousands, upholding justice and fighting against the corrupt imperial government. Historians have confirmed that the story originates from true events which took place during the tumultuous years of the Song Dynasty. Thus the oppressed masses were endeared to the account and it eventually evolved into a folk legend.

Ancient Chinese astrologers believed that in the constellation called Big Dipper there are 36 heavenly stars and 72 baneful stars, each representing a demon. The 108 heroes in *Water Margin* are believed to be the incarnation of the demons represented by these stars.

In this six-volume series, you will find a faithful account of the events leading to the assembly of the outlaws on Liangshan Mountain. Their deeds, which formed the heroic legend of a grand scale, are brought to life through the skilful hand of Singaporean cartoonist Teo Seng Hock.

In this second volume, *Marsh at the Foot of Liangshan*, you will witness why Yang Zhi tried to sell his sword and killed a rogue, how Wu Yong took the convoy of birthday presents by strategy, how the Ruan brothers fought He Tao, and how Lin Chong seized the waterside stronghold out of true fraternity.

We wish to take this opportunity to thank Teo Seng Hock for his vivid illustrations, Wu Jingyu for her translation, and the production team for putting in their best effort in the publication of this book.

Water Margin
Vol 1: Emergence of the Demons
Vol 2: Marsh at the Foot of Liangshan
Vol 3: The Tiger Slayer
Vol 4: The Siege of Zhu Family Village
Vol 5: Confronting the Royal Court
Vol 6: The Grand Assembly

About the Illustrator

Teo Seng Hock 张胜福, born in Singapore in 1965, began working as a cartoonist in 1990 and founded the Shenzao Cartoon Production Company (神造社) in 1993. His works include *City Heroes, Shock Battle, The Cockroach, Stories about Cops*, etc. His books *Sweet Spring* and *Challenger* have drawn large numbers of readers in Taiwan. Teo is now working on the Chinese classic *Water Margin* and hopes that the project will push his creative career to a new height.

About the Translator

Wu Jingyu 吴敬瑜, born in 1928, studied journalism at Yenching University from 1944 to 1948. She studied at the Beijing Foreign Language School in 1950, where translators and interpreters were trained. She began her teaching career in 1954 and has taught Chinese, Chinese Literature, English and European Literature in various American, Canadian and Chinese universities. She is now Professor of English, specialising in teaching English as a second language at the Beijing Second Foreign Language Institute.

Glossary

Liangshan and Liangshan Marsh — Liangshan Mountain is situated in present Liangshan County, Shandong Province. At a height of 197 metres above sea level with precipitous rocks and deep canyons, the mountain is now a famous tourist attraction. In the Song Dynasty, the Liangshan Marsh was a large lake with rough waters and dotted with clusters of reeds and unusual plants. It was situated in the area that is now in the counties of Liangshan, Yuncheng and Juyie.

The pledge 投名状 — The original meaning of the term 'pledge' 投名状 is 'a visiting card', similar to what people use today, on which was written the person's name, official title, status, etc. What Wang Lun demanded from Lin Chong was a human head. It was a rule among robbers in those days to present the head of a man one kills to show his determination to join the band.

A convoy of birthday presents 生辰纲 — In ancient times, goods shipped in large quantity were called 纲, usually using a large number of vehicles or porters. For instance, 盐纲 was a convoy of salt. In the story, Yang Zhi was first in charge of 花石纲, which consisted of exotic flowers, rare herbs and strangely shaped marbles or rocks presented to the emperor. The 生辰纲 is a convoy of birthday presents.

Provincial Commander-in-chief 提辖 — Title of an officer in charge of military training and management in local governments at the prefecture or county level. Lu Da was a 提辖 in the office of the younger Jinglue Zhong. Yang Zhi and Suo Chao were 提辖 in the garrison headquarters at Damingfu.

Arranging the seats — In this book the Liangshan heroes discussed their seating for the first time to decide their relative position in the band. In the entire story of *Water Margin*, the matter of seating was discussed three times. This time Cao Gai was given the seat of honour, that of the Number One leader. Present-day tourists to Liangshan can still see in the Hall of Loyalty and Justice an empty seat beside the chair in which the sculpture of Song Jiang sits. This empty seat is supposed to have belonged to Chao Gai.

Character Introduction

Wu Yong
Nicknamed 'Resourceful Wizard' (the wise star among the Heavenly Spirits). A school teacher in Yuncheng County, he was a learned man and a master of the art of war. He went to Liangshan and became military adviser to the chiefs after joining the others in seizing the convoy of birthday presents.

Bai Sheng
Nicknamed 'Daylight Rat' (the rat star among the baneful stars).

Chao Gai
Nicknamed 'Heavenly King Who Carries a Pagoda on His Palm'. Chief of the Liangshan rebels at one time. Formerly the alderman of East Creek Village in Yuncheng County, he escaped to Liangshanpo when he was wanted for his involvement in the robbery of the convoy of birthday presents.

Ruan the Second
Nicknamed 'Lord Who Stands His Ground' (the dagger star among the heavenly spirits).

Ruan the Fifth
Nicknamed 'Short-lived Second Brother' (the sinning star among the heavenly spirits).

Ruan the Seventh
Nicknamed 'Live King Yama' (the vanquished star of the heavenly spirits).

Zhu Gui
Nicknamed 'Dry Land Alligator' (the prisoner star among the baneful stars).

Gongsun Sheng
Nicknamed 'Dragon in the Clouds' (the idle star among the heavenly spirits). An itinerant Taoist priest and learned in Taoist magic art, he plotted with Chao Gai and the Ruan brothers to seize the birthday presents. He occupied third place in the group at the moment.

Liu Tang
Nicknamed 'Red-haired Devil' (the strange star among the heavenly spirits).

Yang Zhi

Nicknamed 'Blue-faced Beast' (the star of darkness among the heavenly spirits). A descendant of the famous general Marquis Yang Linggong. He passed the imperial military examination and became an officer in the Imperial Guards. He was sent to ship marble and lost the valuable cargo when his ship overturned, and then the convoy of birthday presents under his charge was seized by bandits. He had no way out but to become a bandit himself.

He Tao

Government official in charge of capturing thieves and other suspects.

Wang Lun

Chief of the band of robbers who occupied the Liangshan Marsh before the arrival of Lin Chong and the other rebels. He was a narrow-minded man who was jealous of men more talented and more capable than himself.

Suo Chao

Nicknamed 'Daring Vanguard' (the empty star among the heavenly spirits). An officer in the garrison of the northern capital. He had a contest of force with Yang Zhi, because he was envious of Yang being promoted by Governor Liang.

Lei Heng
Nicknamed 'Winged Tiger' (the retiring star among the heavenly spirits).

Zhu Tong
Nicknamed 'Lord of the Beautiful Beard' (the star of abundance among the heavenly spirits). An inspector in the local government of Yuncheng County. Born in a wealthy family, he liked to make friends with good fellows from the rivers and lakes. He set Chao Gai and Song Jiang free out of personal loyalty.

Deng Long
Second chief of the bandits on the Mountain of Two Dragons.

Song Jiang
Nicknamed 'The Filial and Generous One' (chief star among the heavenly spirits). A clerk in the magistrate's office of Yuncheng County. He got the nickname because he was a filial son and a just man. Being generous, he was always ready to help people in distress. Therefore, he was also called 'Opportune Rain'. He fled to Liangshanpo after he killed a woman who blackmailed him.

Maps of Area around Liangshan during the Northern Song Dynasty

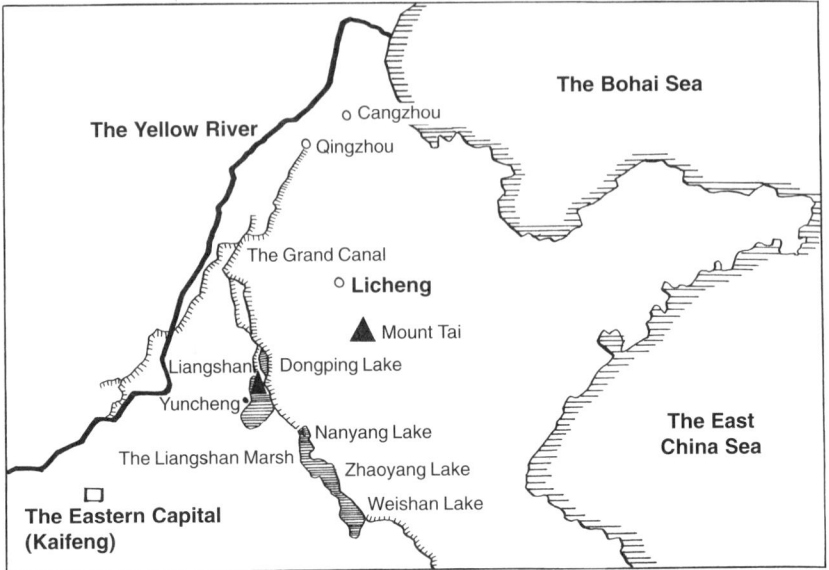

Course of the Yellow River during the Northern Song Dynasty | The Grand Canal | Prefectural Boundaries | - - - - Marshland

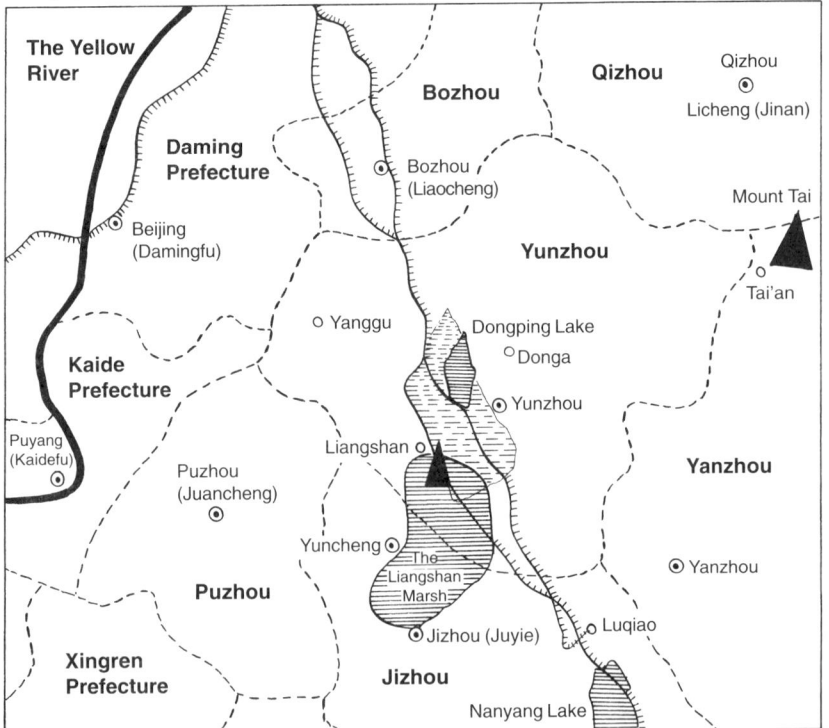

Contents

Appendix
Next Issue
Nicknames of Characters
A Brief Chronology of Chinese History

Recapturing …

The young village head of Shi Family Village, Shi Jin, nicknamed 'Nine-Tattooed Dragons', made friends with the robbers on Shaohua Mountain. When the magistrate was informed of this, he sent troops to surround the village. Shi Jin set fire to the village and escaped to Weizhou, where he came across Lu Da, a major in the district garrison troops.

Lu Da was forthright in character and was always ready to launch out against injustice. One day while having a few drinks with friends in a tavern, he learned that a local despot, a butcher named Zheng who called himself 'Lord of the West', had been tormenting a young woman. Lu Da went to punish Zheng. Their argument led to a fist fight. Lu Da dealt three punches at the butcher and killed him. To avoid being captured by the authorities, he went to Wutai Mountain and became a Buddhist monk. He was given the Buddhist name Zhishen. It was difficult for a man like him to abide by monastic rules. One day in drunkenness he created a riot in the temple. As a result of his riotous behaviour he was sent away by the abbot to Xiangguo Monastery in Dongjing.

In Dongjing, Lu Zhishen met Lin Chong, drill instructor of the 800,000-strong Imperial Guards and they became sworn brothers. Soon, Lin Chong got into trouble because the foster son of Gao Qiu, Commander-in-chief of the Imperial Guards, coveted Lin's beautiful wife. Framed by Gao Qiu, Lin Chong was branded as a criminal and banished to Cangzhou. Yet Gao Qiu still would not leave him alone. He sent his henchmen after Lin Chong, intending to put him out of the way. On the night of a snowstorm, Lin Chong killed his three enemies and fled from Cangzhou …

YANG ZHI
SELLS
SWORD IN
BIANLIANG

Yang Zhi the stalwart Hero
Sold sword in the market and killed a rogue.
Banished to the northern capital,
He proved himself matchless on the parade ground.

Lin Chong walked a whole day in a snowstorm and reached the marsh.

As evening was approaching, it became colder and colder. Lin Chong saw a restaurant by the side of the lake and entered.

Could you tell me how far it is to Liangshan Mountain?

Though it's just a matter of a few miles, you can only get across by boat.

Are there boats for hire?

Those on the mountain are criminals wanted by the government. People here are afraid of getting involved with them. No boatman dares go near the mountain.

Lin Chong was upset when he heard the waiter's words. He decided to have some food first and then see what he could do.

5

He drank several cups of wine. Thinking of his being framed by Gao Qiu, his parting from his wife and his uncertain future, he felt greatly distressed.

Gao Qiu, you damned villain, why do you persecute Lin Chong so?

Oh, how hateful!

Hmm?

Aha! Leopard Head Lin Chong! How thoughtless you are! Don't you know the government has set a high price on your head?

Ah!

OK! Don't be afraid. I was just teasing you, Instructor. Please forgive me.

I am Zhu Gui from the Liangshan Marsh. Wang Lun is our leader. Very pleased to meet you.

At daybreak the next day.

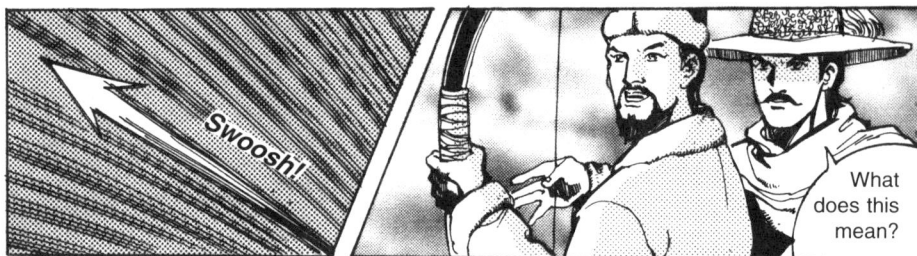

Swoosh!

What does this mean?

This is a signal in our stronghold. As soon as they get the arrow, a boat will be sent over.

In a little while, a boat rowed by a bandit soldier appeared.

Lin Chong followed Zhu Gui into the boat. The boat shot across the lake like an arrow and soon arrived at the Shore of Golden Sand.

A few men were waiting for them.

Turning round the cliff, Lin Chong saw a pass, in front of which were displayed various kinds of weapons.

*"Right wrongs in accordance with heaven's decree."

They walked through two more passes and reached the entrance. The stronghold was surrounded by tall mountains. Within the three sturdy passes on the sides was a piece of level land the size of 1,000 to 1,500 square metres. The main entrance was flanked by many rooms.

This is Lin Chong, drill instructor of the 800,000-strong Imperial Guards in Dongjing. He was framed by Gao Qiu and exiled to Cangzhou. Then the army fodder depot under his care was burnt.

After he had killed his three enemies, he had to escape to the village of Squire Chai Jin. Squire Chai recommended him to join us.

Leader of the band, Wang Lun ...

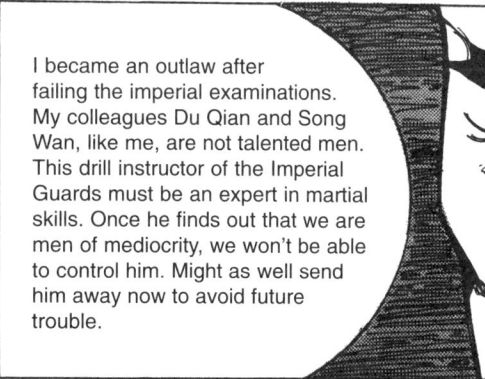

... ...
... ...

I became an outlaw after failing the imperial examinations. My colleagues Du Qian and Song Wan, like me, are not talented men. This drill instructor of the Imperial Guards must be an expert in martial skills. Once he finds out that we are men of mediocrity, we won't be able to control him. Might as well send him away now to avoid future trouble.

11

Instructor, our base is small, weak in military force and lacking provisions. There's not much of a future for you here. We'll give you some silver so that you can find a better place for yourself.

Leader, I hope you understand that Lin Chong did not come to Liangshan to get some silver.

Brother, please don't take offence at what I have to say ...

After all, one more man means more strength, doesn't it?

Lin Chong is a talented man. Moreover, he has the recommendation of Squire Chai, who has been very kind to us. What's going to happen when he hears that we've sent Lin Chong away?

I came to be an outlaw because I committed a capital offence. Why do you suspect me?

You people don't understand. He committed a monstrous crime in Cangzhou. Who knows what motive he has in coming here? Maybe he's a spy.

In that case, you must submit a 'pledge' within three days, to show your determination.

A pledge?!

It's a rule in our band that every man who wants to join us must kill a man and present his head.

That's ...

All right, I'll bring the head of a man I kill within three days.

When Lin Chong joined the outlaws on Liangshan Marsh, Wang Lun, the bandit chief, was afraid that with his superb martial skills Lin would be a threat to his position as the leader. So he demanded that Lin kill a man to show his determination to join the band. Lin Chong went downhill and waited for three days till the first passer-by came along. They fought several rounds, but Lin was unable to subdue him.

With your excellent martial skills, how did you become a bandit?

Please stop fighting!

Let's continue our fight!

Wang Lun gave a feast for Yang Zhi, intending to get him to stay as a counter-balance to Lin Chong.

Gao Qiu is still in office. I don't think he tolerates men of talent like you.

As Yang was determined to get an official post in Dongjing, Wang Lun failed to persuade him to stay.

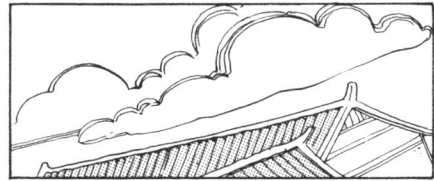

Yang Zhi left Liangshan the next day and reached Dongjing after some days.

I'll be very grateful for your help.

Yang bribed officials in the Privy Council. His request to be reinstated was finally granted.

At the cost of lots of silver, he had an audience with Gao Qiu.

Nine out of ten lieutenants sent to bring marble here came back. You're the only one who lost a cargo and ran away. Though you've been pardoned of the offence, I can hardly give you another job.

Having spent all his money to bribe the Privy Council officials, Yang Zhi became penniless. The only thing of value in his possession was a sword he had inherited from his ancestors.

He had no alternative but to sell the sword.

Precious sword for sale!

Suddenly, he noticed people in the street running in confusion to hide themselves.

It turned out that they were running away from a bully called Niu Er, whose nickname was 'Hairless Tiger'.

How much do you want for this sword, man?

Three thousand taels. It's a precious sword handed down by my ancestors.

*祖傳寶刀一把

* *Precious sword for sale!*

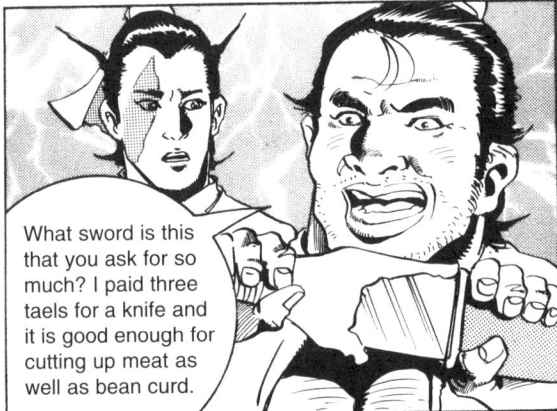

What sword is this that you ask for so much? I paid three taels for a knife and it is good enough for cutting up meat as well as bean curd.

Why don't you try out my sword, if you don't believe me?

Actually Niu Er had no intention to buy the sword. He was just harassing Yang Zhi.

Yang Zhi lost his temper.

Niu Er went down the bridge and demanded 20 coins from someone.

OK. Can you cut this stack of coins in half?!

My sword can cut iron or bronze without twisting its blade and kill a man without getting blood stains on it. When a hair is blown against the edge it's cut in two. You just open your eyes and look!

Dang!

Wahhhh!

Wahhhh!

Though it does cut the hair, I don't believe it can kill a man without getting blood stained. I don't believe you until you kill someone to prove it.

I don't have to kill a man. I can demonstrate by killing a dog.

You said kill a man, not a dog. Are you trying to talk your way out of it?

You have no intention to buy my sword. Why do you keep pestering me?

Why don't you chop me with your sword, if you have the guts?

Boom!

Bang!

How dare you push me!

Whooh!

Hoosh!

I have killed the rascal, and I'll take all responsibility. I just want you to come along as witnesses to the magistrate's office.

In court Yang Zhi related how he came to kill Niu Er and the others testified that he had told the truth. On account of his having voluntarily surrendered himself, the prefect spared him the caning and ordered him to be detained for the time being.

Knowing that he had rid the town of a bully, the jailers all respected him.

After two months, Yang Zhi was caned 20 strokes and then banished to the northern capital Damingfu.

34

The wealthy people in the neighbourhood collected some money and presented it to Yang Zhi. They also begged the two guards escorting him to take good care of him on the way. The guards agreed.

When they arrived at Damingfu, the guards delivered the report about Yang Zhi from the Kaifeng prefecture to the garrison headquarters.

Grand Secretary Liang — Commander in chief of the garrison troops in Damingfu. He was the son-in-law of the emperor's tutor Cai Jing.

You seem to be a man of ability, Yang Zhi, just the kind of man I need. I'll give you a job in my office.

I am very grateful for your kindness, Your Excellency.

Grand Secretary Liang was very pleased that Yang Zhi proved to be diligent and attentive to his duties. He intended to give him a commission in the army.

36

Dong!

Dong!

Everything is ready, Your Excellency.

Liang had a mind to promote Yang Zhi, but he was afraid that other officers may resent it. He ordered a general parade and inspection of the troops to be held on the drill ground to give Yang Zhi a chance to show off his martial skills.

Zhou Jing, lieutenant to commander-in-chief Liang

Boom!

With the spikes removed and lime rubbed on the cloth wrapped ends of the spear, Yang Zhi and Zhou Jing fenced.

The number of white spots on each contestant's black garment will decide who is the winner!

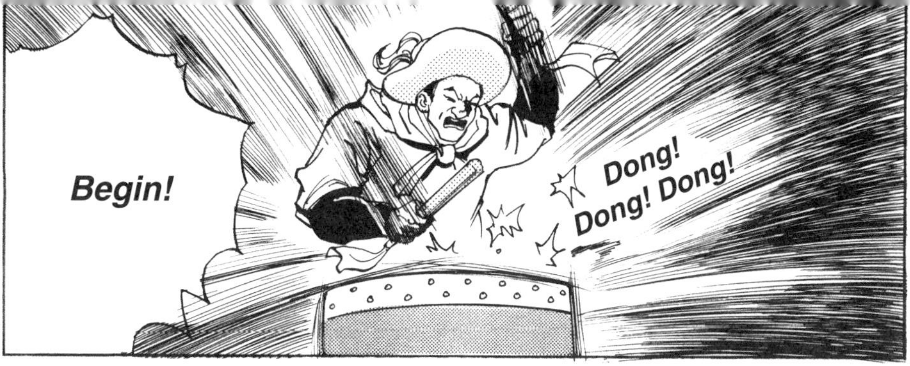
Begin!
Dong!
Dong! Dong!

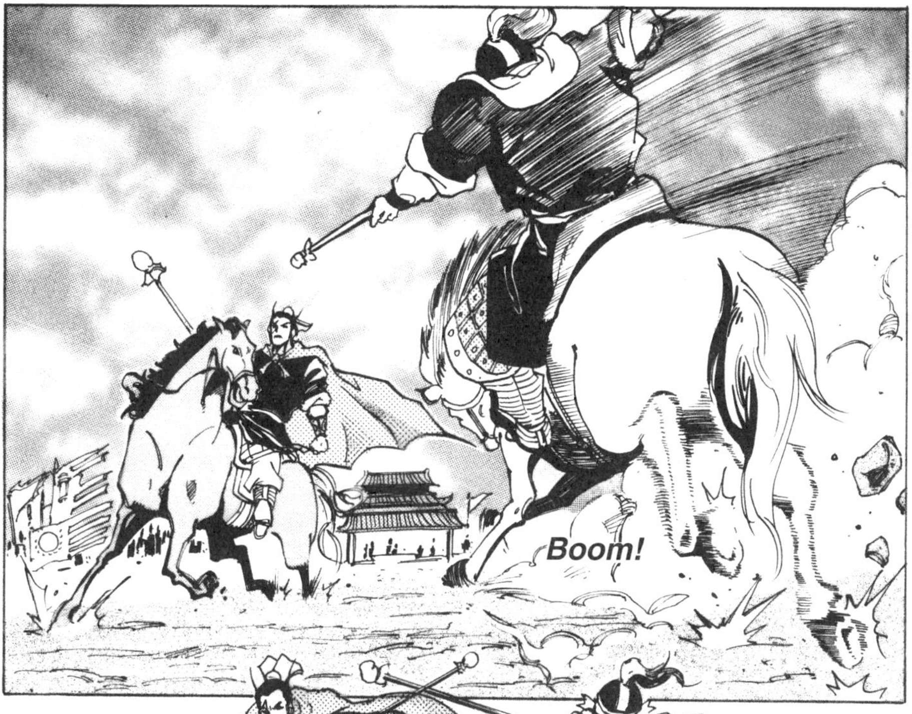
Boom!

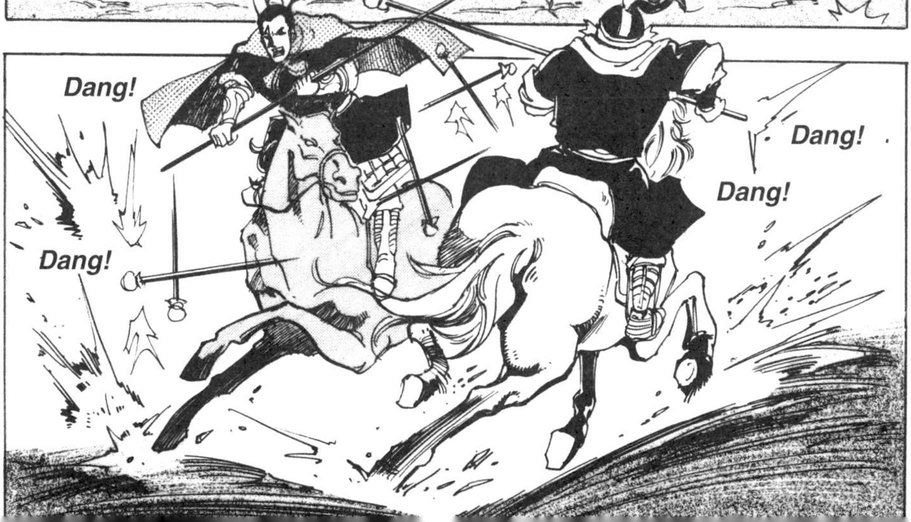
Dang!
Dang!
Dang!
Dang!

Thank you, sir!

There were 30 to 50 white spots on Zhou's garment, but only one under the left shoulder of Yang's.

The other officers may not be convinced. Perhaps ...

You two should now contest as archers.

His Excellency is quite determined to see me win.

Yes, sir!

Ah!

They put arrows to their bows. Suddenly, Yang Zhi turned round with his back towards Zhou Jing.

Shoot! You can have three shots at my back.

41

The third
shot!

Shoosh!

Boom!

Ah!

Now it's my turn.

Waw! Waw!

Superb!

Aiya!

I have no enmity towards him, I won't shoot any vital spot.

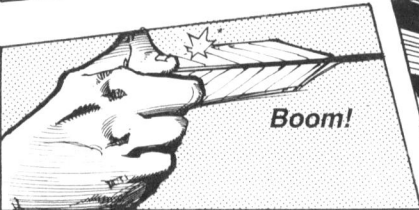

Boom!

Wah!

Arrggh!

With such excellent skills, Yang Zhi should take over the official position Zhou Jing has been deprived of.

Thank you for the promotion, Your Excellency!

Wait!

Your Excellency, if he defeats me, he can have my position instead of Zhou Jing's.

46

The man was Captain Suo Chao, nicknamed 'Daring Vanguard'. He was Zhou Jing's master.

I think ...

I think it's a good idea. Let Yang Zhi compete with Suo Chao, then the other officers will see exactly who is better.

All right. Yang Zhi you can have a contest with Captain Suo Chao.

Captain Suo, I'm learning from you.

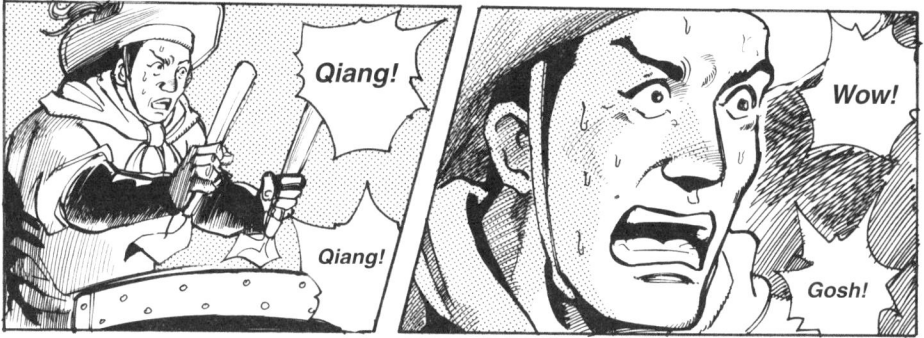

Well done!

My goodness!

They have fought more than 50 rounds and it's still hard to say who'll be the winner.

Both of them are valiant warriors. It would be a pity if either of them was hurt.

Your Excellency, both Suo Chao and Yang Zhi have proven their excellent military skills. I suggest that you order them to stop, so that no one will be injured.

His Excellency has decided to promote both of you to the rank of major.

Thank you, Your Excellency!

Yang Zhi and Suo Chao retired to change their clothes and then returned to the platform to thank the Grand Secretary. They also saluted each other as they formally accepted their commission.

Ti! Ta! Ti! Ta!

**WU YONG TAKES THE CONVOY OF
BIRTHDAY PRESENTS BY STRATEGY**

Men like tigers, horses like dragons,
Ingenious strategy is used on Ridge of Yellow Earth.
Loads of gold and jewel brought to the stronghold
To the great annoyance of the Grand Secretary.

Grand Secretary Liang sent runners from his office to extort gold, pearls and other valuables from the common people to be sent as birthday presents to his father-in-law, the royal tutor Cai Jing.

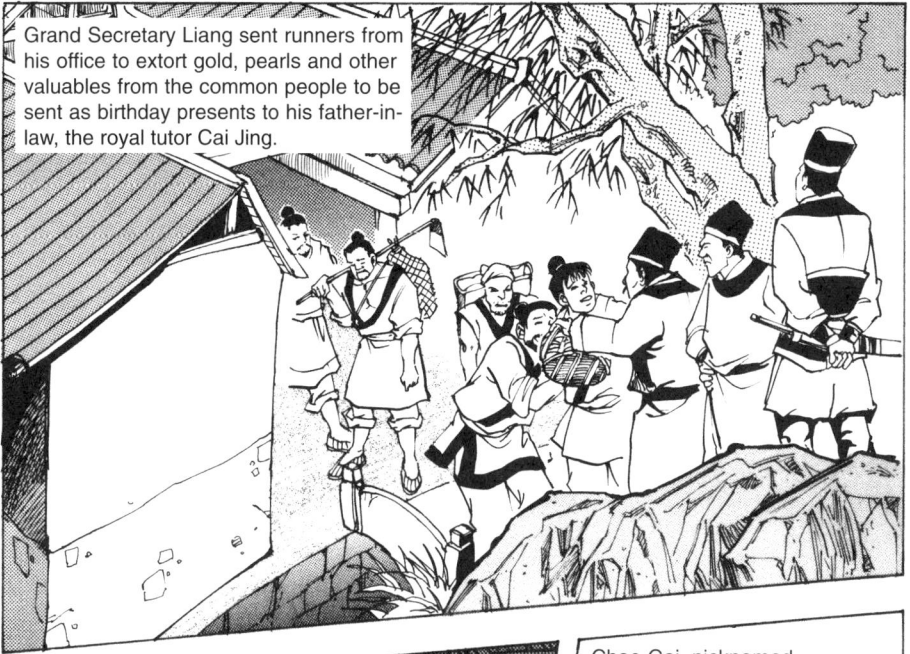

The alderman of East Creek Village in Yuncheng County.

Chao Gai, nicknamed 'Heavenly King Who Carries a Pagoda in His Palm'.

A gregarious man, he liked particularly to make friends with brave men, and was always ready to help the poor and needy. That day, he was discussing a plan with the village scholar Wu Yong and a man called Liu Tang.

Sir, do you think we can do it? It is ill-gotten wealth, extorted from the common people. I don't see why we shouldn't take it.

Wu Yong, 'Resourceful Wizard'.

If Brother Chao intends to do it, I will assist you.

Sir, being a man of unusual wit, you must have suggestions to make?

This is something that takes seven or eight like-minded people to accomplish.

I know three good and reliable men in Shijie Village.

The eldest is called Ruan the Second, 'Lord Who Stands His Ground'; the second brother is Ruan the Fifth, 'Short-lived Second Brother' and the third is Ruan the Seventh, 'Live King Yama'.

East Creek Village in Yuncheng County ...

I learned that Grand Secretary Liang in the northern capital is sending birthday presents worth 100,000 taels of cash to his father-in-law Cai Jing in Dongjing. The convoy will pass this way via Yellow Earth Ridge.

There's a man in East Anle Village near Yellow Earth Ridge whom I once helped. I think we can stop at his place and get him to help us.

Liu Tang

Gongsun Sheng

Chao Gai

We can take the birthday presents only by strategy. Here's what I think we'll do.

The Ruan brothers, the Fifth, the Second and the Seventh.

We are to disguise as ...

Ha, ha! Great! No wonder people call you 'Resourceful Wizard'. You are indeed comparable with Zhuge Liang.

Ha, ha ...

One day in June at the Yellow Earth Ridge ...

Hurry up!

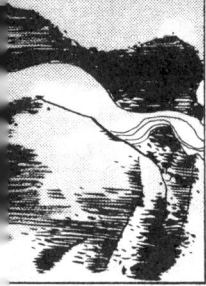

Hurry up! You can have some rest after we get over the ridge.

Waw! It's really hot. The stones on the road are burning my feet.

The heat is really unbearable! I've got to sit down for a rest.

We've got to hurry past the ridge. Then, I'll let you have a rest.

Huh!?

Let me find out who these people are. You watch over the soldiers.

Those people are moving furtively. Could they be bandits?

Ah!

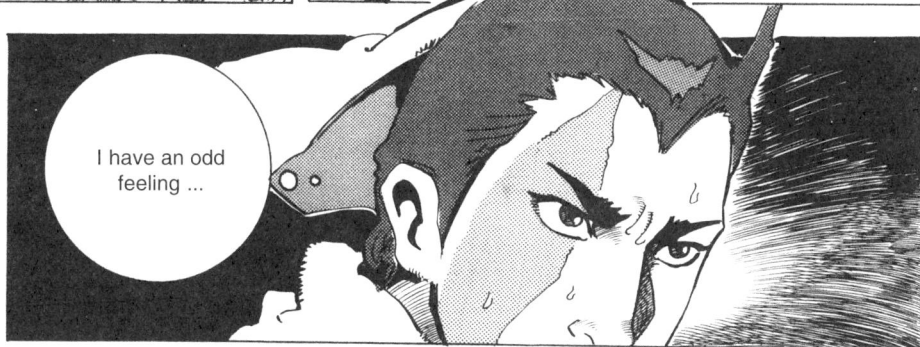

Burp!

Mmm, good wine. It's really quenched my thirst!

There's not a drop of water on the ridge. Please allow the soldiers to buy the wine, Major Yang.

The date traders have drunk a bucket and nothing has happened to them. The soldiers really need something to assuage their thirst.

Mmm ...

Since you say so, you can tell them to go ahead, but they must hurry up. We've got to be on our way without further delay.

Hey! Sell us the other bucket of wine!

No, no. My wine is drugged. I won't sell it to you.

We were just joking. Don't take it seriously.

Ah!

Oh, my!
The wine is drugged!

Ha, ha, ha, ha ...

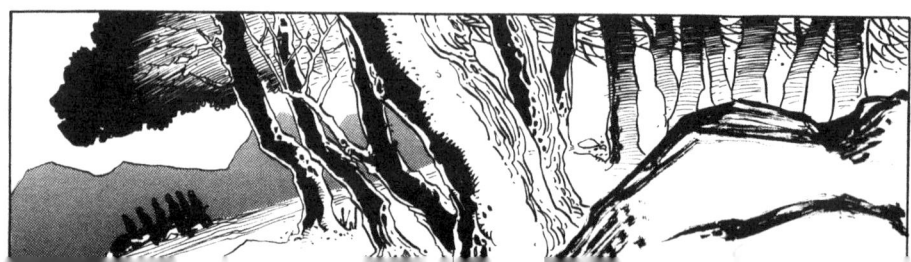

I'm in serious trouble!

You people have done me in!

I cannot face the Grand Secretary again, now that I've lost the birthday presents. But where can I go? It seems the Liangshan Marsh is the only place.

My God!

What a disaster! What am I to say when I get back? I should have listened to Yang Zhi.

Mmm, where is he?

Steward, Major Yang has left us. What are we to do?

Yang Zhi is gone, and now ... We might as well ...

I've got an idea. We can put all the blame on him.

Eek!

Huh!

Where are
you bastard
from?

Swoosh!

He's a
formidable
opponent.

Blonk!

Halt! Who are you?

I am Yang Zhi, nicknamed 'Blue-faced Beast'.

The one who tried to sell a sword in Dongjing and killed a man called Niu Er?

Yes, that's who I am, and may I ask who you are?

'Flower Monk' Lu Zhishen.

Weren't you living in the Xiangguo Monastery in Dongjing? Why are you here?

I saved Lin Chong and infuriated Gao Qiu who ordered the Xiangguo Monastery to drive me out. Then I heard that the Precious Gem Temple was a safe place to go.

But the chief Deng Long refused to let me join his band of robbers. After I defeated him in a fight, he retreated to his stronghold and barred all mountain passes leading to it. I failed to break my way into the stronghold, so I came to this forest to take a rest.

A friend of mine runs a wine shop near here. Let's go there and chat over a few drinks.

Sorry, I mistook you for one of the gang and attacked you.

Good idea. A few drinks are just what I need.

That's the place.

If you intend to go to the Two Dragons Mountain, I think you can do it this way ...

Excellent plan!

Deng Long, chief of the bandits on Two Dragons Mountain.

Good news, Chief! The monk has been captured.

Really! Wonderful! Bring him in.

This fat monk said he was going to get the bandits on Liangshan Marsh to attack you ...

So we got him drunk and then tied him up to bring him to you as a prisoner.

You've never anticipated coming to such an end, have you, wicked monk?

Come, take him outside and chop him into pieces!

Chief!

Let me handle them!

Whump!

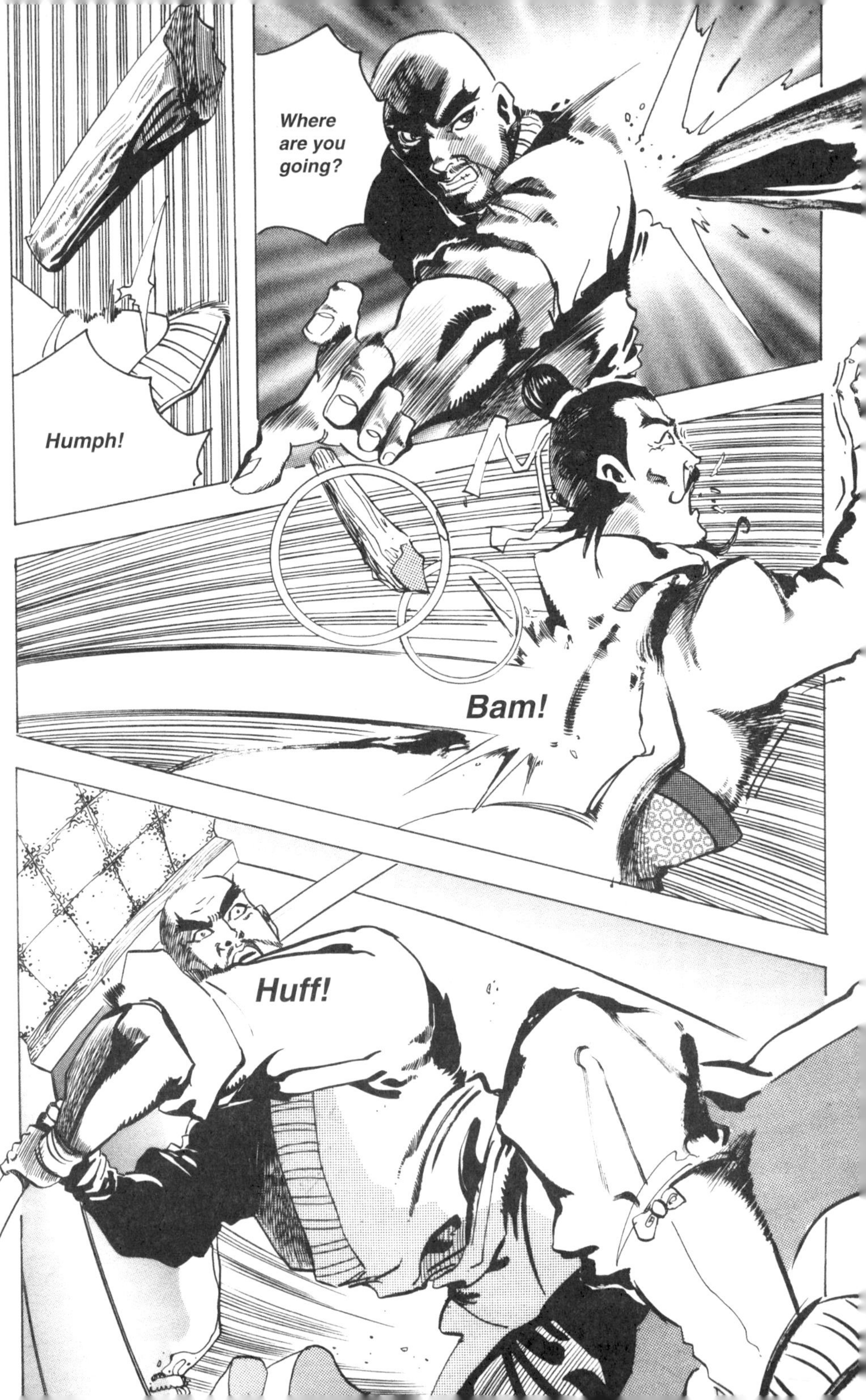
Where are you going?
Humph!
Bam!
Huff!

Go to hell!

Bam!

Chief!

Please spare us, heroes!

Being men of superb military skills, you should be masters of our stronghold.

Those of you who want to stay can be sure of a good future. Those who don't want to follow us can get the hell out of here immediately. OK? Bring some wine!

* Two Dragons
Mountain

THE RUAN BROTHERS
FIGHT HE TAO

The sly bandits and their chief sneaked away,
The pursuit fails as the noise alerted the fugitives.
A piece of casual writing rouses
Warriors of heavenly and baneful stars

The steward and soldiers who escorted the convoy of birthday presents returned to Beijing. When reporting to Grand Secretary Liang, they put all the blame on Yang Zhi. The furious Grand Secretary sent a dispatch to the prefect of Jizhou ordering him to arrest Yang and others involved in the case.

Clip Clop!

He Tao, Chief thief catcher at Jizhou Prefecture.

He Tao led some of his men to the Yellow Earth Ridge, but he found no trace of the bandits. He returned home in exasperation.

He Qing — He Tao's younger brother.

I have all those bandits in my pocket here. Hee, hee!

Elder Brother, don't be so upset.

What? How can you have the bandits in your pocket?

Don't you see the names written on this piece of paper? They are the bandits'.

It turned out that He Qing was a gambler. On the third day of the sixth month, he copied the register from an inn-keeper. Among the merchants who had stayed at the inn were several date traders. He Qing noticed that their leader was Chao Gai.

I guess what happened on the Yellow Earth Ridge was Chao Gai's doing. You only have to seize Bai Sheng, then you'll find out who the others were.

Wonderful! He Qing, come with me to the office. We'll go and arrest Bai Sheng.

That night He Tao led eight thief catchers to Anle Village.

He Qing asked the innkeeper to take them to Bai Sheng's home and to get him to open the door.

Is Bai Sheng home? I'm Wang, the innkeeper.

Who is it?

I'm the proprietor of the inn.

Bai Sheng has gone on a trip.

What is it, Mr. Wang?

He's left some things at my place. I've brought them to him.

Ah!

What ...
what do
you want?

Come!
Take him!

Bai Sheng! We
have proof that
the robbery at the
Yellow Earth
Ridge was
committed by
you!

Your Honour!
I'm innocent! I
know nothing
about the
robbery.

Don't try to
fool us. We'll
make you
talk! Come
on!

Bind the woman!

Your Honour, I really don't know what happened on the Yellow Earth Ridge. You're accusing the wrong person.

On the third day of the sixth month, I copied the register at the inn in Anle Village. Seven date traders stayed there for the night. You were seen selling wine the next day.

But I'm really innocent. I had been sick for quite a few days at the time.

Since he refuses to talk, let's search!

The runners noticed that one spot under the bed was not level. So they started to dig. When they got to three feet deep, a bundle of gold and jewel was found.

Humph! What's this? How are you going to explain this?

That's our hard-earned savings.

Your leader is Chao Gai of the East Creek Village. Right? Tell the truth!

We know for sure that you're one of the bandits. Who are the other six?

Your Honour, I am indeed innocent. I have nothing to tell you, even if you beat me to death.

The prefect sent He Tao to Yuncheng County to arrest Chao Gai and the other bandits.

Despite the torture, Bai Sheng refused to plead guilty.

I'll go there early when the office is open.

I'm going to the office to get the arrest warrant. Wait for me here.

Mmm?

It's very quiet in there. Maybe it's past the morning office hour.

Hey, waiter, do you know who's on duty at the office today?

Here comes the officer on duty.

Song Jiang, nicknamed 'Opportune Rain', a scribe in the magistrate's office of Yuncheng County.

Sir, I am a thief catcher from the Prefecture of Jizhou. May I ask your name?

I am Song Jiang.

Could you tell me what your official business is?

Chao Gai, the alderman of East Creek Village under the jurisdiction of your county, is the leader of a group of seven robbers who made off with the convoy of birthday presents ...

Chao Gai is a trusted friend of mine. Now he's involved in such a monstrous crime. If I don't save him, he's probably going to lose his life.

This is a matter of great importance. You re to deliver the warrant sonally to the magistrate. t now the morning office over and the magistrate is taking a rest, you'll have to wait. I am also going home.

ou're quite t, sir. Don't t me keep u from your usiness.

I've got to tip off Brother Chao about it.

Why are you in such a hurry, sir?

I came at the risk of my own life to tell you that the robbery on Yellow Earth Ridge has been traced to you.

Please take care, Brother, and go away with your three friends as quickly as possible.

Of the 36 stratagems, the best is running away. Let's first seek refuge in Shijie Village near Liangshan Mountain. If the government troops come hot on our heels, we can join the band on Liangshan.

He Tao and his men surrounded Chao Gai's village. Suddenly they saw a blaze rising inside. The sky was filled with soaring flames and black smoke.

Where are you criminals running?!

Fire!

Let's make a getaway from the back!

Lei Heng - 'Winged Tiger'.

Brother Zhu Tong, the alderman has left the village safely.

Fine!

Zhu Tong - 'Lord of the Beautiful Beard'.

The fire burned fiercely. After chasing the fugitives the whole night without catching anyone, He Tao decided to withdraw his soldiers.

Write down the confession of these two farmers, so that Investigator can submit a report to the prefect.

He Tao rode back to Jizhou with the official report.

Good! Now I bid you take 500 troops and a thief-catcher investigator to arrest the criminals in Shijie Village.

The Ruan family home in Shijie Village.

Government troops are coming towards our village!

These damned soldiers are courting their own deaths!

Mr. Wu, take the birthday presents and your family to the Lijia intersection.

The rest of us will stay here to deal with the troops.

Humph! Let's interrogate a few fishermen to find out their whereabouts.

Yes, Sir!

There's no one in Ruan the Second's home.

The fishermen say that the other two Ruan brothers live on the lake. The robbers must be hiding on the lake too.

Give my orders! Get in your boats. We're to seize the robbers on the lake!

In more than 100 boats the government troops raced towards the home of Ruan the Fifth.

All my life I've been fishing in Liao'er Lowlands, Growing neither hemp nor paddies. When all corrupt officials are wiped out, I'll have demonstrated my loyalty to the Song Dynasty!
There's a boat in the marsh, Sir.
You damned government lackeys, how dare you come here to provoke me?! You're stroking the tiger's whiskers backwards!
'Short-lived Second Brother', Ruan the Fifth.
He is Ruan the Fifth!
Close in on him!

Shoot!

Boom!

Go after him!

Suddenly whistles were heard from the reed marsh.

This one is Ruan the Seventh.

Having grown up in Shijie Village, I am by nature inclined to kill. My very first target is He Tao, the investigator, Whose head I will present to the emperor.

Ruan the Seventh, 'Live King Yama'.

Hurry up! Capture the robber!

Boom!

Boom!

It turned out that the fleet of
boats, each two tied together,
with no crew and loaded with dry
reeds and faggot, was aflame.
Now the boats were blown by the
wind and drifting towards them.

Waw!

Burrr ...

Ya-a-a!

Boom!

The soldiers who were lucky enough to swim ashore were killed by the Ruan brothers, Chao Gai and Gongsun Sheng.

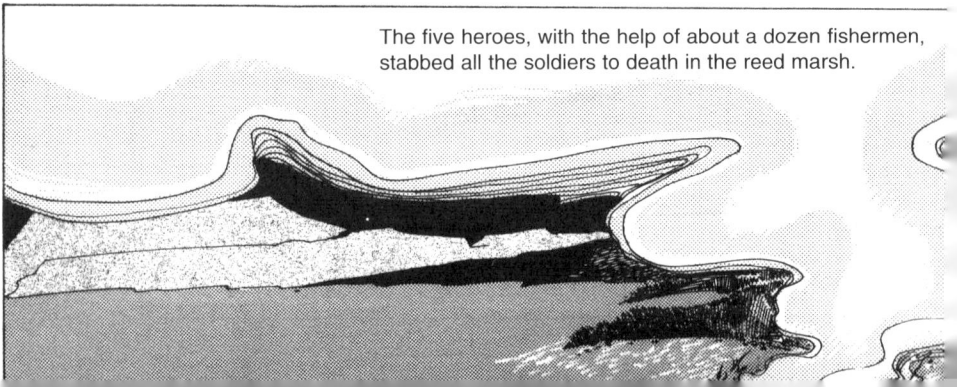

The five heroes, with the help of about a dozen fishermen, stabbed all the soldiers to death in the reed marsh.

Please have pity on me, Sir. I have an 80-year-old mother.

We were going to chop your head off at one stroke ...

Now we'll let you off. Go back to tell the bastard prefect of Jizhou that we Ruan brothers and Heavenly King Chao Gai of East Creek Village are not to be trifled with.

Even if Cai Jing came in person, we'd poke dozens of holes in his body.

We are sparing your life and teaching you a lesson.

Yes, yes, yes, Sirs!

Ruan the Seventh took out a dagger, sliced off He Tao's ears, and then released him.

LIN CHONG SEIZES THE WATERSIDE STRONGHOLD

The deceitful Wang Lun is put to death,
And sagacious Chao Gai now in command.
Lin Chong turns against the headman out of true fraternity,
And proves his unequalled lofty quality.

Hearing about their arrival, Wang Lun and the other leaders went downhill to greet them.

Chao Gai, Wu Yong, Gongsun Sheng, Liu Tang, the Ruan brothers and their families as well as valuables reached the stronghold on Liangshan.

At the banquet Chao Gai gave their hosts a brief account of how they took the convoy of birthday presents and how they defeated the pursuing troops afterwards.

Hall of Fraternity

Wang Lun gave a banquet in the Hall of Fraternity to welcome them.

Wow! These are no ordinary people. If I let them stay, they'll be a threat to my own position!

What we did is a serious offence against the law. We are deeply grateful to Wang Lun for granting us asylum.

Wang Lun doesn't seem to me to be an open-hearted person. If he meant to let us join the band, he'd have discussed at the banquet our positions in the band.

You're quite right.

It seems that Lin Chong was indignant over Wang's behaviour. There may be a dispute between them.

I've heard a lot about you, Instructor Lin. Could you tell me who recommended you to the leaders here?

Squire Chai Jin.

You mean Chai Jin, 'Little Whirlwind'?

Anyone recommended by Squire Chai Jin should be given the first place in the leadership. That's an accepted principle among good fellows of the rivers and lakes.

I joined the band voluntarily. I don't care whether I'm placed high or low. However, Wang Lun is a narrow-minded man, very difficult to get along with.

Please don't be suspicious, heroes. I came here to let you know that you can count on me to deal with him if he should utter anything against you.

If Wang Lun is reluctant to let us stay, we'll find shelter elsewhere.

Soon, a messenger came from Wang Lun to invite Chao Gai and his comrades to attend the banquet in the pavilion on the south side of the mountain.

Wang Lun is an example for whoever contests my action!

Hereafter we'll obey your orders.

I killed him not to seize his position.

Brother Chao Gai is generous in aiding needy people, and he is both brave and resourceful. He is the man to lead us. Don't you agree? And, I also suggest that we appoint Mr. Wu Yong as his adviser.

Yes!

Thus, Lin Chong made Chao Gai take the seat in the centre. Flanking him were Wu Yong, Gongsun Sheng, Lin Chong, Liu Tang, Ruan the Second, Ruan the Fifth, Ruan the Seventh, Du Qian, Song Wan and Zhu Gui, in that order.

Today Instructor Lin appoints me to be the chief of the stronghold and Mr. Wu Yong to be my adviser. From now on we will run the stronghold together, each of us having his own duties: garrison the stronghold, store up provisions and fodder, etc. We'll make a concerted effort to accomplish our just cause.

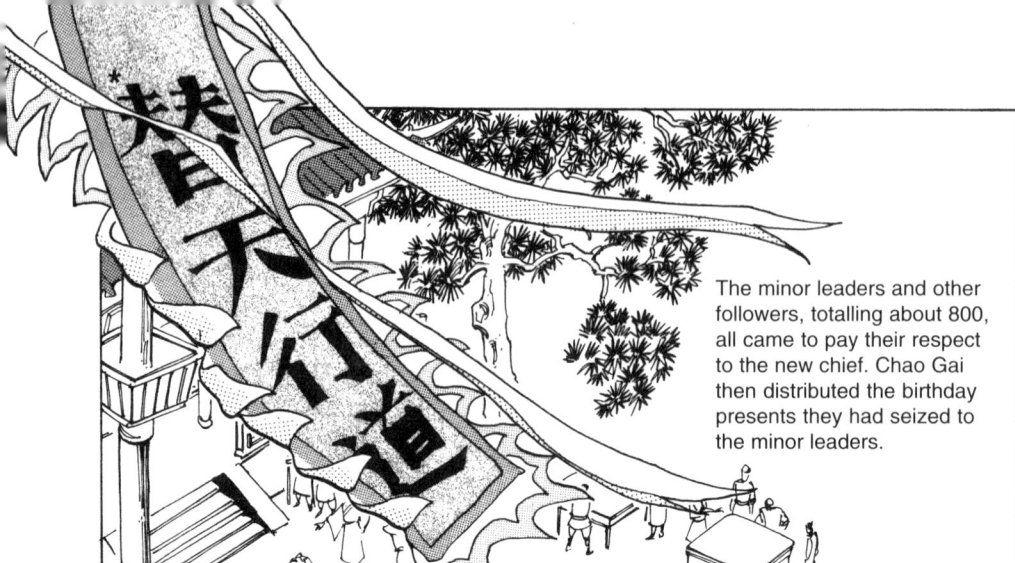

The minor leaders and other followers, totalling about 800, all came to pay their respect to the new chief. Chao Gai then distributed the birthday presents they had seized to the minor leaders.

* Right Wrongs in Accordance with Heaven's Decree

They spent the money that was left to consolidate the defence works and embankments, make weapons and armours and helmets. Chao Gai encouraged everyone to make concerted efforts to safeguard their stronghold.

After that they drilled the men in fighting on water in preparation for an attack from the government troops.

In Jizhou, He Tao who escaped from Shijie Village reported to the prefect how the troops he led were routed. Then they learned that Chao Gai had become the new chief on Liangshan.

The prefect dispatched drill master Huang An with 1,000 troops to attack Liangshan Marsh.

The troops collected boats from the surrounding villages and moved towards the marsh by two routes. They vowed to wipe out all the Liangshan bandits.

139

After receiving the news Chao Gai discussed with his adviser Wu Yong how to deal with them.

A scout reported that the prefect of Jizhou had sent troops to attack their stronghold.

We must give them a heavy blow in the battle on the lake, so that they will think twice before coming near us again. I have discussed our military moves with adviser Wu Yong. You are all to obey his orders.

The Ruan brothers.

Each of you take ten men and sail out in a boat, and ....

Then Wu Yong summoned Lin Chong and Liu Tang and told them what they were to do. He also gave orders to Du Qian and Song Wan. Thus the rebels were well prepared for the attack.

Huang An and his 1,000 troops reached the Beach of Golden Sand. They saw no one on the lake, but heard the sound of battle horns.

Look, Your Honour!

The Ruan
brothers!

Shoot!
Kill them
all!

Wah la la ...

Swoosh!

Swoosh!

Swish!

After them! Hurry up! Don't let anyone escape!

No, Your Honour, we mustn't pursue them!

Our troops in East Creek Village were tricked by the bandits Du Qian and Song Wan and got stranded at the narrow mouth of the creek. The officers and soldiers were attacked by the enemy who surrounded them. They had to abandon their boats and run for their lives.

What?!
The troops on land have been wiped out by Lin Chong, and all their horses were seized by the bandits.
Ah!
We've underestimated our enemy!
Give my orders! Turn round the boats and stop the pursuit!

All troops withdraw on land!

After hearing the report Huang An waved a white flag and signalled his troops to retreat.

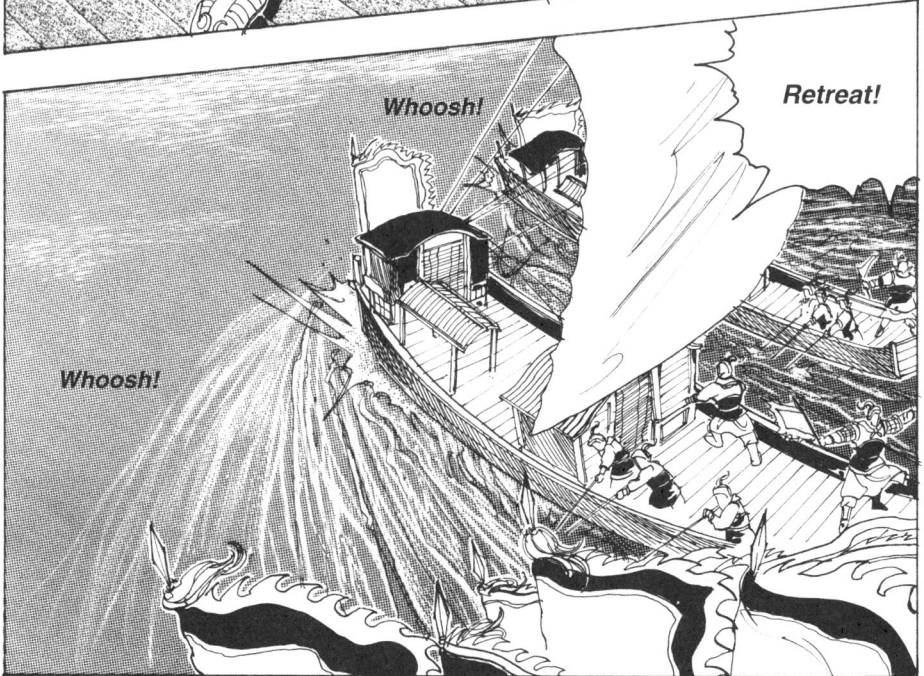

Whoosh!

Whoosh!

Retreat!

145

Huang An! Where can you escape?

Ah!

Whoosh!

Boom!

Splosh!

Ha, ha! Now you know who you are facing, don't you?

The Liangshan leaders returned to the headquarters in triumph. They had seized 600 good horses and a large number of boats and weapons. They also captured alive about 200 enemy soldiers.

Lin Chong, Du Qian, Song Wan, the Ruan brothers and Liu Tang were commended for their outstanding performance in battle and the soldiers were rewarded.

梁山水軍

梁山水

*Liangshan Navy

After Chao Gai and his colleagues formed the new leadership of Liangshan, they sent Liu Tang to deliver 100 ounces of gold to Song Jiang as a token of their gratitude to him for saving their lives.

It takes a lot of courage for you to come here!

Brother Chao Gai is now the leader of the rebels on Liangshan. He sent me here to deliver to you this letter and 100 ounces of gold.

I cannot accept the gold. You need it to build up your stronghold. You should go back as soon as possible and not linger here.

Please convey my greetings to your leaders.

Where are you heading, sir?

Oh, it's Mrs. Yen.

You must come today t spend some time with my daughter.

Song Jiang had once helped this woman whose name was Yen. To thank him, she made him take her daughter Poxi as a concubine.

When the girl was carrying on with his assistant Zhang Wenyuan, he stayed away from her home.

Please pay no heed to rumours. If my daughter has any fault, I'll punish her.

Daughter, your Third Brother is here!

Don't be so headstrong. I've brought him here, be nice to him!

It's getting late, you two should go to bed.

My daughter is young and spoiled. Please have patience with her.

Humph! The slut has gone to sleep as if I were not here. I'll lie down, have some rest and leave as soon as day breaks.

About five o'clock, Song Jiang got up.

He left the house to go home.

Morning, sir. Come and have a bowl of steaming hot soup.

Oh, it's Old Man Huang.

Give me back that letter!
So you've been having illicit communication with the bandits. I'm going to report to the magistrate!
Give that letter to me. I'll agree to all your terms.
I have three conditions: One, you allow me to marry Zhang Wenyuan; two, I'll keep all the jewellery; and three, give me all the gold.
I agree to the first two conditions, but not the third. Because I did not accept the gold. Give me my bag first and allow me a few days to get some money to give you.
No. Give me the money and then I'll give you the bag.
You slut!

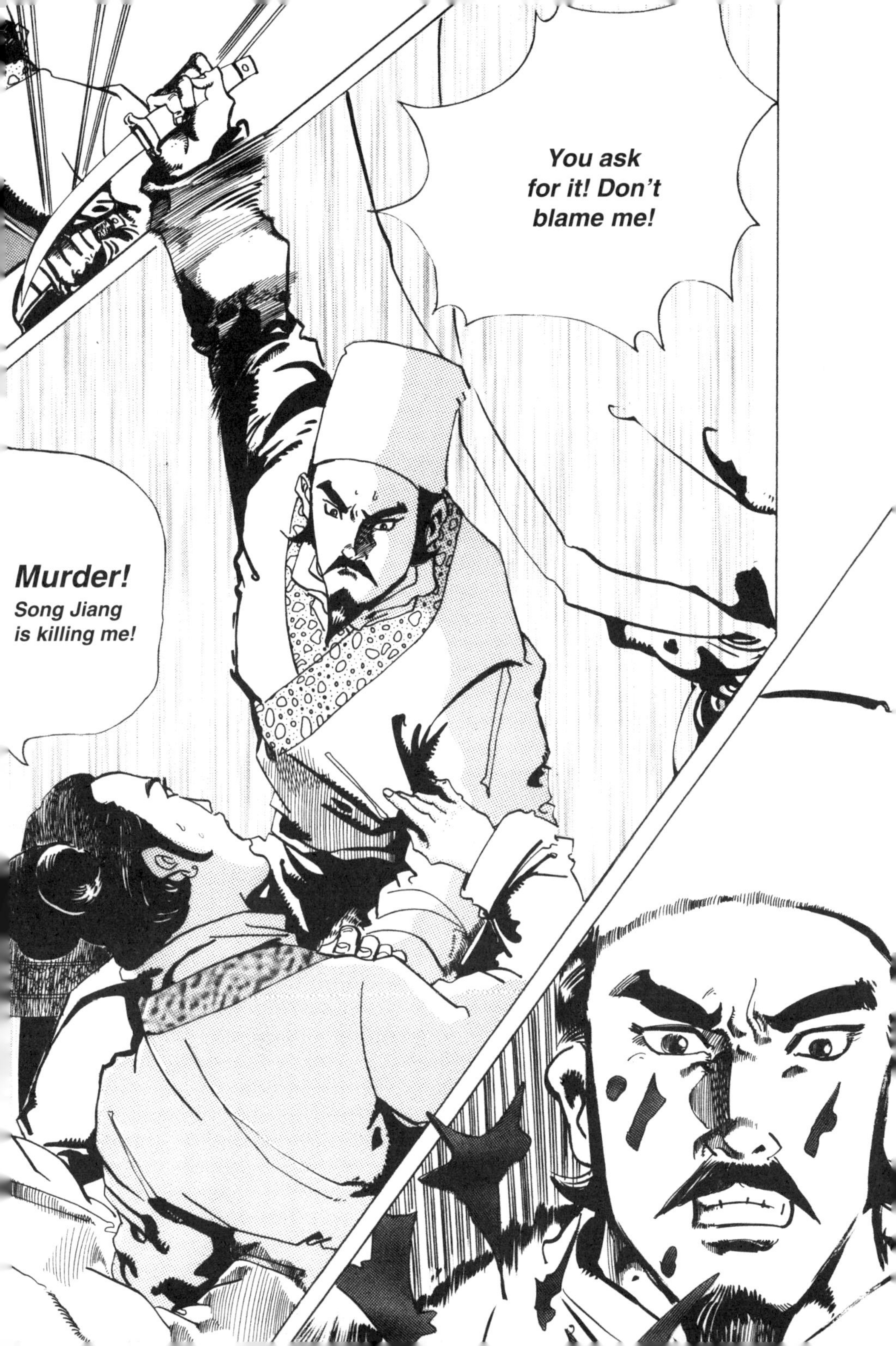
You ask for it! Don't blame me!
Murder! Song Jiang is killing me!

After Mrs. Yen reported the murder, the magistrate sent Zhu Tong and Lei Heng to arrest him. Being Song Jiang's friends, Zhu and Lei helped him flee his home village to seek shelter with Chai Jin in Cangzhou.

Next Issue

The escape of Song Jiang leads to the story of Wu Song, the Tiger Slayer. After killing the tiger, Wu Song avenges the murder of his elder brother. And then, in righteous fury, he gets rid of a despot and his lackeys. The story gets more and more exciting, and one hero after another make their appearances. Please go on to read the next volume *Water Margin — The Tiger Slayer*.

Nicknames of Characters

**The Filial and Generous One
呼保义,
Song Jiang**

I have round eyes like brass saucers and a murderous look between my brows. I was born a trouble maker. Lions shiver with fear and vipers are stricken with terror when they see me. People who offend me are bound to be short-lived. You can have a try if you don't believe what I say!

Short-lived Second Brother, Ruan the Fifth

My nickname means: "Hail fellows and defend justice." If people called me King or Lord Song Jiang, the fraternal affection between us would be lost. My mother called me Black Third Brother because of my dark complexion. But that address is not for you to use. Well, you can call me Opportune Rain. I am a most generous person, always ready to help others. Just tell me what difficulties you have and I will give you timely help as rain on parched soil!

I have a bristly chin and an awesome bearing. What I hate most are cowards who care for nothing but saving their skin. In battle, the soldier who charges at the head of the troops is no one but me!.

I was the alderman of East Creek, in charge of the civil affairs of my village. Once a stupid priest built a pagoda with grey stone and drove the devils from West Creek Village to mine. In a fit of fury I picked up the pagoda and carried it on my palm across the river.

Heavenly King Who Carries a Pagoda in His Palm, Chao Gai

Daring Vanguard, Suo Chao

I have delicate features, fair skin and a long beard. I am well-read in classics and versed in the art of war. People call me 'Resourceful Wizard'. Do you suppose Zhuge Liang is any better than me?

I am called a Blue-faced Beast just because there is a blue-coloured birthmark on my face. The 'beast' is not a wild brute but a fierce animal, say, like a lion or a tiger.

**Blue-faced Beast,
Yang Zhi**

**Resourceful Wizard,
Wu Yong**

It is not exaggeration to call me a winged tiger. I am seven and a half feet tall and an expert in martial skills. I can leap over a wall and jump across a 20 to 30-feet-wide stream as easily as a swallow does. It's a pity that there was no Guinness Book of World Records in my time.

Everyone says I look like Lord Guan, the God of War. Actually, I am much more handsome than he was. I am eight and half feet in height with a reddish brown complexion. I have a pair of eyes sparkling like stars and a beard one and a half feet long. You say that Lord Guan's beard was two feet long? But a shorter one is now in fashion.

**Lord of the Beautiful Beard,
Zhu Tong**

**Winged Tiger,
Lei Heng**

I have bulging eyes and pimples on my face. There are dark patches on my body, as if I were cast in iron or bronze. I am the very image of King Yama.

I am a Taoist priest very good at Taoist magic. I can command the wind and rain, and ride on mists and clouds. I am tall and impressive-looking. 'Dragon in the Clouds' is a very suitable name for me!

Dragon in the Clouds, Gongsun Sheng

Live King Yama, Ruan the Seventh

Do I look like a devil? I'm called one just because there are some hairs on the red birthmark on my temple. My face is wide and dark red, my hefty body, too, is covered with black hair. If I were to live in modern times, I'd probably be called a red-haired "foreign devil"!

I was born with a protruding chin, a wide mouth, bushy eyebrows and a cold gleam in my eyes. I got the nickname because I look exactly like the devil 'Lord Who Stands His Ground'.

Lord Who Stands His Ground, Ruan the Second

Red-haired Devil, Liu Tang

《亚太漫画系列》

水浒传
水泊梁山

原作：施耐庵
绘画：张胜福
翻译：吴敬瑜

亚太图书有限公司出版